T0012109

A Study Guide for
Baby Dinosaurs on the Ark?

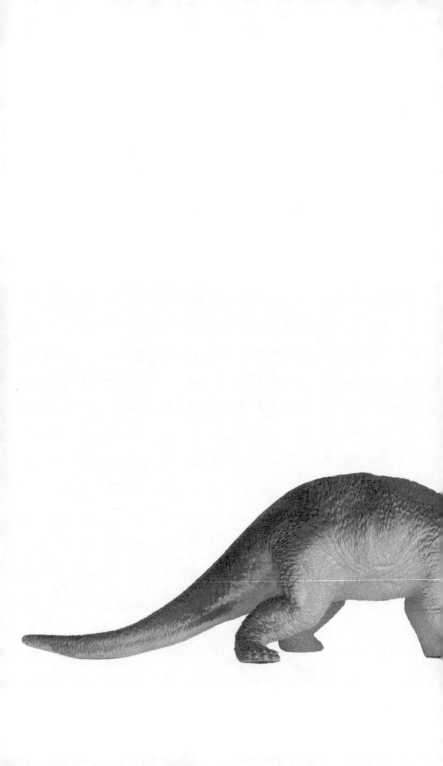

A Study Guide for
Baby Dinosaurs on the Ark?

JANET KELLOGG RAY

WILLIAM B. EERDMANS PUBLISHING COMPANY
GRAND RAPIDS, MICHIGAN

Wm. B. Eerdmans Publishing Co.
4035 Park East Court SE, Grand Rapids, Michigan 49546
www.eerdmans.com

29 28 27 26 25 24 23   1 2 3 4 5 6 7

ISBN 978-0-8028-8269-1

Library of Congress Cataloging-in-Publication Data

A catalog record for this book is available from the Library
   of Congress.

Contents

Introduction to the Study Guide

When you're told evolution and faith aren't compatible, the options are bleak: you can reject a vast body of science evidence, or you can reject God.

Baby Dinosaurs on the Ark? The Bible and Modern Science and the Trouble of Making It All Fit explores ways in which people of faith force-fit modern science into Genesis in an attempt to sustain a literal reading of the Bible.

The foundations of evolution theory—genetics, comparative anatomy, chemistry, physics—get very complicated, very fast. *Baby Dinosaurs on the Ark?* is written for the non-scientist, and specifically, the non-scientist with questions about creationist explanations for scientific evidence.

Baby Dinosaurs on the Ark? invites those with questions into a conversation:

Are you uncomfortable when faith leaders tell you that scientists are wrong about the age of the earth and the development of life?

Do you find yourself rethinking, deconstructing, or reconstructing religious teachings in which you were raised?

Are you searching for ways to maintain both your faith and your intellectual honesty?

Do you simply wish you knew more about the science supporting evolution and the age of the earth?

This study guide is designed for use in small groups, in a classroom, or by independent learners. The guide supports a discussion leader who may not be well-versed in the science of origins or the perceived conflicts between faith and science.

The study guide incorporates many ways of engaging the material:

- summarizing facts and content
- understanding, comparing, and explaining topics
- analyzing and critiquing ideas presented in the book
- examining motivations and applications of concepts
- defending personal positions
- connecting concepts to ideas beyond the scope of the book

I encourage you to journal as you read and study. This can be as conventional as writing in a separate notebook, creating a computer document, or simply making quick entries on your phone's notes app. Formally or informally, use the study guide to engage the content.

Whether you study and journal privately or discuss in community, let the study guide be a companion as you address hard questions about creation, science, and the Bible. Read. Think. Tackle the questions head-on. Dare to form a faith that doesn't ignore the questions.

Here's what you'll find in the study guide:

Chapter Highlights

Read or reread each book chapter before beginning each chapter's study. Chapter highlights are listed and summarized to help you recall the content of each chapter.

Discussion

The discussion prompts engage multiple levels of thinking. Prompts guide recall, understanding, analysis, and evaluation of chapter content.

Digging Deeper

Each study guide chapter includes a section in which a topic is expanded beyond what is presented in the book.

Resources

Each study guide chapter includes a list of resources that can be used to introduce the chapter or to explore a topic in more depth. Resources include short videos, documentaries, links to online readings, books, and interactive learning experiences.

1

The Biology Professor Who Doesn't Believe in Science

Chapter Highlights

- When you're told evolution and faith aren't compatible, the options are bleak: you can reject a vast body of science, or you can reject God.
- A young earth and a literal creation week are the default positions for many religious traditions—anything else is considered atheism.
- The theory of evolution says nothing about God or religion or any other worldview.
- Belief and acceptance are different—the foundation of one is faith; the foundation of the other is empirical evidence.
- Without a fundamental knowledge of evolution theory and opportunities to ask questions and express doubts, it is often difficult to accept evolution.
- Acceptance of evolution is often a journey beginning with acceptance of an ancient earth.
- Science doesn't have an answer for every question.

Discussion

1. In what ways does popular culture imply that faith and science are at odds?

2. Compare belief and acceptance. What are some things you accept?

3. Several common personal beliefs are identified in this chapter: the love of family, the incarnation, and the resurrection. What are your most dearly held beliefs? Can your beliefs be "proven" using the scientific method? Why or why not? If not, does this discredit your beliefs?

4. Is evolution synonymous with atheism? Why might someone hold this position?

5. Confirmation bias occurs when our desires directly influence our beliefs. If we really *want* something to be true, we may believe it is *actually* true. As a result, we tend to only accept sources that confirm what we already believe and disregard any facts to the contrary. Do you see confirmation bias in the faith/science debate?

6. The "David versus Goliath" motif is common in both religious and secular circles. We often take personal pride in holding a minority opinion against any position deemed more powerful or popular. The religious tradition in which I was raised wore its uniqueness from other groups as a badge of pride. Have you ever worn a minority position, either religiously or secularly, as a badge of pride? Can you identify any pitfalls in doing so?

7. What would it look like in your life to "love God with heart, soul, *and* mind"?

8. Have you ever felt forced to choose between faith and science? If so, what were the circumstances? How did you feel in the midst of a faith/science tug-of-war?

9. Is the credibility of faith at stake in the faith and science discussion? Why or why not?

10. After reading the foreword and chapter 1, what are your thoughts regarding acceptance of evolution and belief in a personal and loving God?

Digging Deeper: Galloping Duane Gish

For decades after the Scopes Monkey Trial, creationists were relatively silent about teaching creationism in public-school classrooms. Not about to let the Russians and their Sputnik satellite win the race to space, the United States poured a fortune into science education through the National Defense Education Act in the late 1950s. This act funded new science textbooks, and the teaching of evolution got a shot in the arm.

Scientists generally avoided debates with creationists, not wishing to legitimize creationism as an alternative to evolution. But by the late 1970s, bills were popping up in state assemblies calling for equal treatment of "creation science" and "evolution science" in public schools. In 1981, the National Academy of Sciences and the National Association of Biology Teachers sounded an alarm.[1]

The situation was declared a crisis in American science education.

Enter a silver-tongued orator: a champion of the creationists who presented himself in the mold of Galileo, facing down the powerful scientific establishment.

The orator was Duane Gish.

Gish was known as "creation's bulldog" by the anti-evolution crowd, a nod to Thomas Henry Huxley, called "Darwin's bulldog" in the nineteenth century.[2] A biochemist by training, Gish held several academic positions until he became associated with the Creation Research Society and later the Institute for Creation Research.

As the star of the creationist debate circuit, Gish participated in hundreds of debates, often on college campuses, across the United States and around the world.

Gish relished his reputation as a fighter on the creationist-evolution stage. Instead of a tightly formatted debate with limited topics, Gish preferred a no-holds-barred approach with speakers given forty-five minutes or more of uninterrupted time.

Gish used his stage time to overwhelm his opponents with a barrage of questions and statements that could never be individually addressed and refuted, even within an hours-long timeframe. Anthropologist Eugenie Scott coined the term "Gish Gallop" to describe a debate approach in which a speaker overwhelms the opponent with a multitude of arguments, with no regard for accuracy or relevance.[3]

Gish inevitably pointed out all the things his pro-evolution opponent failed to address, and debate victory was always declared.

The Gish Gallop is a technique now recognized in debates across a wide variety of topics. In a climate of "I did my own research," the Gallop can be recognized whenever someone makes generalized or unsubstantiated claims, tells anecdotes with little or no value, misrepresents facts, or refutes statements no one has actually made.[4]

Resources

David Masci, "For Darwin Day, 6 Facts about the Evolution Debate," Pew Research Center, February 11, 2019, https://www.pewresearch.org/fact-tank/2019/02/11/darwin-day/.

> From the Pew Research Center. A quick breakdown about evolution acceptance in the United States, by demographics. This article also contains many handy

links to resources about evolution acceptance in both the United States and other countries, creationism in American public schools, and acceptance of evolution by scientists.

BioLogos, https://biologos.org.

Bookmark this site now! An invaluable resource, this website features world-class scientists and theologians, videos, articles, events, and other resources, all affirming both faith in God and acceptance of modern science. Founded by Dr. Francis Collins, former head of the National Institutes of Health and winner of the Templeton Prize for "harnessing the power of the sciences to explore the deepest questions of the universe and humankind's place and purpose within it."[5] Physicist Deborah Haarsma is the current president of BioLogos.

Francis S. Collins, comp., *Belief: Readings on the Reason for Faith* (San Francisco: HarperOne, 2010).

A beautiful collection of essays on faith from a who's who of influential writers, speakers, and world leaders: Augustine, Locke, Pascal, L'Engle, Tutu, Wiesel, Polkinghorne, Mother Teresa, and many more.

2

Making Science Fit Genesis

Chapter Highlights

- When science is portrayed as the enemy, it can be difficult to love God with heart, soul, *and* mind.
- For many people of faith, an outright rejection of science feels wrong, especially in the twenty-first century.
- From cosmology to geology to dinosaurs, Christians have attempted to reconcile science evidence with a literal reading of Genesis. Usually, the solution was straightforward—the science was retrofitted and forced into Scripture.
- An alarming number of Americans no longer believe they can be a person of faith and accept the science of evolution.
- On one end of the spectrum, the loudest voices claim that a person of faith cannot accept evolution. On the opposite end of the spectrum, the loudest voices claim that a valid scientist cannot be a person of faith.
- Movies, television, and other forms of popular culture perpetuate the idea of a war between faith and science.

Discussion

1. Within communities of faith, science is often portrayed as an enemy of faith, especially in the context of origins. Are there other contexts in which people of faith might see science as an enemy?

2. "Baby dinosaurs on the ark" is used as a metaphor for all the ways we try to force-fit science into the Bible, without respect for the ancient cultural context, literary genre, or the writer's original intent. Is this your experience? Can you identify other "baby dinosaurs" forced upon the Bible?

3. Why do many people of faith feel the need to make science fit Genesis? Is this a good, bad, or neutral goal?

4. At one time, the Ptolemaic earth-centered model of the solar system was the best science of the day. With the advent of better instruments and better mathematics, the earth-centered model was replaced with the (correct) sun-centered model. The best science at one point in time can be replaced with a more current, better science. Does this trouble you?

5. How are people of faith typically portrayed in popular media? Have you seen any of the *God's Not Dead* movies? In your opinion, do these movies present a realistic portrayal of the faith/science dilemma? Explain.

6. While there are plenty of scientists who are not people of faith, scientists like Richard Dawkins who overtly have an axe to grind against religion are rare. Interestingly, Dawkins and famous young earth creationist Ken Ham agree: evolution and Christianity are incompatible. Why do these two voices at the extremes of the spectrum command the most attention?

7. There is a "how to send your student off to college" cur-

riculum titled "Welcome to the War" published by a popular young earth creationist publishing house. What impact might this approach have on a Christian student at the outset of their higher education?

8. Which is (or was) harder for you to accept—an old earth or the common ancestry of all life? Why might this be so?

9. Do you personally struggle with conflict between your religious beliefs and science? How do you (or did you) deal with the conflict? Is ignoring or not thinking about the tension a strategy?

10. Do you know anyone who abandoned faith (or is thinking of doing so) because it doesn't fit with science?

Digging Deeper: The Problem of Extinction

You might say the two fishermen caught more than they bargained for.

The year was 1666. Two local fishermen in a boat off the coast of Italy caught a giant shark, which was subsequently hauled in and beaten to death (being pre-PETA). They lopped off the head and sent it to famous Dutch naturalist Nicolas Steno.

While examining the shark's head, Steno experienced an "a-ha" moment: the shark teeth looked exactly like common rocks found everywhere in the hills of Tuscany. The locals simply called them "tongue-stones."

Conventional wisdom said that fossils were nothing more than curiously shaped rocks. Steno proposed something radical—these "rocks" were actually teeth from once-living sharks! Others had suggested that fossil "rocks" had living origins, most notably Leonardo da Vinci and science virtuoso Robert Hooke. Steno, however, took things further. He proposed a means of fossil formation from living tissue and explained how fossils of creatures like sharks could be found in rocks far from the sea.

By the eighteenth century, fossils were accepted as formerly living things. This idea, however, was mighty troubling to many.

French anatomist Georges Cuvier rose to fame in the early nineteenth century. So skilled was he, Cuvier could reconstruct an entire animal with only a few fragments of bone.

Cuvier was particularly puzzled by fossils of elephants found near Paris. These elephants were distinctly different from any known living elephant, but were elephants, nonetheless. Some proposed that these giant distinctive creatures still lived but were playing a clever game of hide-and-seek somewhere on the planet. Cuvier wasn't buying it.[1]

As more and more massive mammalian fossils were discovered, Cuvier suspected that maybe, just maybe, things on earth were different in the past. Periodically, proposed Cuvier, catastrophic events led to widespread destruction of species.

Cuvier established extinction of species as a fact to be reckoned with.

"Extinctions?" cried many naturalists. "Say it isn't so!"

Everyone knows that God created each creature specially, according to a Divine Plan. How can it be that massive numbers of creatures perished? How can it be that fossil creatures are so different from living creatures? Surely these animals must exist in some unexplored part of the world!

Darwin was not the first to suggest that species had changed over time. Transmutation of species, an idea proposed by Jean Baptiste Lamarck, preceded Darwin's theory of evolution by natural selection.

But Cuvier was no evolutionist. Although Cuvier's landmark discoveries about extinction would prove to be essential to Darwin's work, Cuvier adamantly opposed the concept.

Instead, Cuvier described a dark and dangerous time before Eden and before Noah's flood. This mysterious time in history was the source of the many rock layers filled with extinct creatures. In this scenario, God created and destroyed a pre-Genesis world that preceded the biblical seven-day creation.

Following Cuvier's lead, other geologists and paleontologists expanded the idea of pre-Eden creations. As fossil formations were identified and found to be consistent across Europe, the puzzle intensified. In 1842, Alcide d'Orbigny characterized ten different Jurassic formations as ten separate creations, prior to Eden.

But it didn't stop there. More discoveries, more strata identifications, more creations. When all was said and done, European geologists recognized twenty-seven separate creations and floods, all occurring before the Genesis creation story.[2]

Eventually, things got so complicated that geologists and paleontologists gave up. It was impossible to reconcile a literal Genesis with the science observed in the fossil layers. Far from jettisoning the Bible, devout geologists and paleontologists instead just shrugged their shoulders and admitted that the Bible cannot explain the observed science.

Resources

April Maskiewicz Cordero, "The E Word," filmed April 2016 at
       TEDxPointLomaNazareneUniversity, Point Loma, CA,
       YouTube video, 17:34, https://youtu.be/tttop-dO8XQ.

> Dr. Maskiewicz Cordero is a professor of biology at
> Point Loma Nazarene University. Surveys given during
> the first week of her evolution course find that more
> than 90 percent of her students reject evolution as the
> explanation for how humans came to be on earth. In
> this TEDx talk on evolution and faith, she identifies
> three myths that must be overcome if Christians are to
> embrace or be open-minded about evolution.

"A Cosmic Adventure through Science and Faith," Science for
       Youth Ministry, 2017, http://scienceym.org.

> A series of four videos, in a fun Bill Nye (the Science
> Guy) style. Created with a grant from the Templeton
> Foundation, the videos are downloadable and free, and
> they are accompanied by student and leader guides.
> Videos are hosted by astrophysicist Dr. Paul Wallace.

"Layers of Time," American Museum of Natural History,
       https://www.amnh.org/explore/ology/paleontology
       /layers-of-time2.

> "Layers of Time" is an interactive fossil game. Choose
> your level: easy, medium, or hard. After a quick intro-
> duction to fossils, put rock layers in order from oldest
> to youngest. Get hints and check your work!

3

What Is Science?
The Nature of Science and EVO 101

Chapter Highlights

- Evolution is often dismissed as "only" a theory.
- Science theories like germ theory, molecular theory, gravitation theory, and evolution theory are not unsupported educated guesses. Theories provide scaffolding in a field of science and give meaning to facts and laws.
- Science is truth-seeking. Science seeks to know what is.
- Darwin was not the first to suggest that life had evolved, but he was the first to suggest a mechanism by which it happened: natural selection over vast periods of time.
- Darwin was the first to suggest common ancestry of all life.
- Natural selection does not result in perfection. Instead, a species is "fit enough" if it can survive, find a mate, and leave offspring.
- Evolution is often derided as a random process, but evolution by natural selection is not random.
- Natural selection can only act on the genes already present in a population.

Discussion

1. What do you think when you hear the term "theory"?
   Were you surprised to learn that evolution theory holds
   the same rank as germ theory, gravitation theory, and
   molecular theory?

2. Science is truth-seeking. Science wants to know what
   can be known. Are you challenged or comforted by
   these statements?

3. Rare is the person who questions germ theory, gravitation theory, or molecular theory. Yet evolution theory
   is frequently questioned. Why might this be so?

4. What most people who question or reject evolution
   know about evolution comes from anti-evolution
   apologetics sources. How would you judge your level
   of knowledge/understanding of the theory of evolution prior to reading this chapter? From what kinds of
   sources did you gain your knowledge?

5. Science works to explain evidence without a goal of
   "winning." Why is this important?

6. Scientists are always trying to prove each other wrong.
   A good scientist will try to disprove her own work. Do
   you find this reassuring or not? Explain.

7. What are some limitations of science? Will we ever have
   a complete understanding of the natural world?

8. In his book *The Death of Expertise*, Tom Nichols writes
   about an American climate in which "all things are
   knowable and every opinion on any subject is as good
   as any other."[1] Do we trust experts in general, even if we
   do not understand the details of the field? Specifically,
   do we trust experts in science?

9. A common anti-evolution argument goes something like
   this: in a room full of monkeys banging on typewriters,

what are the chances that a monkey will produce the complete works of William Shakespeare? The chances of such a random occurrence are so small as to be impossible. Likewise, the argument goes, the chances of random mutations resulting in the breadth of life on earth are impossible. Therefore, evolution is impossible. How is this analogy a misunderstanding of how evolution works?

10.   Environmental pressures are a powerful driver of evolution. Before you read this chapter, how would you have explained the physical similarities of very different animals like the shark (a fish) and the dolphin (a mammal)? How would you explain the physical similarities between the marsupial animals of Australia and the distantly related placental mammals found elsewhere on the planet?

Digging Deeper:
Superbugs—Evolution Explains It All

I love a fortuitous accident—especially in science (see chapter 5 for one of my favorites). I love when scientists find what they were *not* looking for but were smart enough to know they'd found it.

In the not too distant past, childbirth, food poisoning, getting a tooth pulled, or even a simple cut could be a death sentence. In World War I, one-third of deaths were due not to weaponry but to infection. In the 1920s we had woefully few options for treating bacterial infections.

It was early fall in London, 1928. Scottish medical researcher Alexander Fleming was anticipating two weeks in the country with his family and was anxious to get out of the lab and on with his holiday.

An accidental hero to all who maintain an unstructured (read: messy) desk, Dr. Fleming left stacks of petri dishes piled on his lab bench. Fleming opened a window and shut the door to his lab for the duration. Left behind were haphazard petri dishes, all inoculated with nasty boil-causing, disease-causing staphylococcus bacteria.

After his holiday, Fleming noticed that one of the dishes was growing a mold—likely carried by a breeze through the open window. His next observation changed the course of human history.

While most of the dishes were covered with the staph bacteria, the dish growing mold was not. In fact, the moldy dish had a clear "kill zone" surrounding the mold where no bacteria grew.

Fleming hypothesized that the mold was producing some sort of substance that killed the bacteria. Fleming dubbed his new discovery a very un-science-y sounding term—"mold juice." Later, Fleming renamed the bacteria-killing substance after the type of mold he found growing in the dish: penicillin.

For the next decade, little was done to develop penicillin as a drug for humans. Still, penicillin was recognized as a magic bullet against many bacterial infections. So precious and rare was this new antibiotic, the urine of treated patients was filtered to retrieve any leftover functional penicillin.

In 1945 the Nobel Prize in Physiology or Medicine was awarded to Fleming and two others responsible for developing penicillin as a widely available drug. Fleming, always modest, said, "Penicillin started as a chance observation. My only merit is that I did not neglect the observation."[2]

But taking a more ominous tone, Fleming then sounded the alarm about overuse of this magic bullet. Fleming predicted that overuse of penicillin would result in penicillin-resistant bacteria.

By the 1950s penicillin-resistant bacteria were commonplace. Today, antibiotic-resistant bacteria ("superbugs") cause more than 2.8 million infections in the United States each year. How do such simple and tiny organisms manage to outwit the best of pharmaceutical science?

Evolution explains it all.

It starts with a mutation. Mutations are copying errors made when DNA is replicated prior to cell division. Reproduction in bacteria is not complicated—a bacterial cell simply divides in two.

A mutation can be beneficial, harmful, or have no effect at all. But this is important: the bacteria do not "try" to mutate in a helpful way. Mutations are simply random copying errors.

Mutations happen by chance, but whether a mutation takes off in a population of bacteria is *not* random.

Penicillin kills bacteria by destroying the cell walls of bacteria. If a random mutation occurred in a species of bacteria that strengthened the cell wall of those bacteria, that species would be resistant to anything attempting to destroy the cell wall.

If these mutated bacteria were growing in the absence of penicillin, we would never know the mutation existed. The cell-wall-strengthening mutation would be of no advantage in the absence of penicillin. The mutation could be passed along for generations but never take off in the larger population.

But in the presence of penicillin, the mutated bacteria would be at a definite advantage. The only bacterial cells capable of surviving the onslaught of penicillin would be the cells with the mutation. All other cells, lacking the mutation, would die.

The survivors pass along the mutation to their off-spring. Before long, the entire population of this bacterial species would carry the penicillin-resistance mutation.

So we develop new antibiotics. But somewhere lurking in the bacterial population may be a mutation resistant to the new approach. It's an ongoing arms race.

Quite simply, antibiotic resistance is evolution by natural selection.

Environmental pressures drive evolution. In the absence of penicillin, resistance provides no advantage. When penicillin is added to the environment, resistance becomes an advantage. Only bacteria with the mutation survive. The bacterial population evolves.

And natural selection can only act on the genetic toolkit already present in the population. Bacteria do not mutate because they are exposed to penicillin (or any other antibiotic). Research demonstrates that resistant bacteria had the resistance *before* exposure to an antibiotic, not as a result of the exposure.[3]

Bacteria provide a real-time, front-row seat to the process of evolution. Bacteria have short generation times and large populations, and because they are always dividing, they are always mutating. Bacteria allow us to witness evolution at warp speed. Evolution works the same in more complex organisms, but we seldom see it in real time.

Overuse of antibiotics and failure to complete a course of antibiotics contribute to the antibiotic resistance problem, but evolution explains *why* we see what we see. As geneticist Theodosius Dobzhansky famously said, "Nothing in biology makes sense except in light of evolution."

Resources

"The Making of a Theory: Darwin, Wallace, and Natural Selection," HHMI BioInteractive, August 26, 2014, video, 31:02, https://www.biointeractive.org/classroom-re sources/origin-species-making-theory.

> Part dramatized documentary and part narrative by evolutionary biologist Sean B. Carroll, this short film tells the story of the voyages of Charles Darwin and Alfred Russell Wallace and how they independently arrived at the theory of evolution by natural selection.

Understanding Evolution, https://evolution.berkeley.edu/evo lution-101/.

> Evo 101 is a treasure of evolution resources from the University of California–Berkeley. Search a multitude of topics and find concise, easy-to-understand explanations, student resources, teacher resources, and links to related topics.

"Natural Selection and the Rock Pocket Mouse," HHMI Bio-Interactive, August 26, 2014, video, 10:31, https://www .biointeractive.org/classroom-resources/making-fittest -natural-selection-and-adaptation.

> This film explores a classic case of natural selection and how the environment drives evolution.

"Discovering the Great Tree of Life," Yale University, January 21, 2009, YouTube video, 11:38, https://youtu.be/mD94D0KAn2U.

> With excellent graphics and beautiful video, this short film from the Yale Peabody Museum of Natural History is an introduction to the interconnectedness of all living things.

Michael Greshco, "Turned to Stone," *National Geographic*, June 2017, 92–105.

> This *National Geographic* article includes beautiful photos of the best-preserved dinosaur of its kind ever found.

4

Where Are You Camping?
A Look at Beliefs

Chapter Highlights

- Young earth creationism, old earth creationism, intelligent design, theistic evolution/evolutionary creationists, and naturalism/scientism describe the primary "camps" regarding religion, evolution, and the age of the earth.
- The terms *creationist* and *creationism* are applied to a specific approach to Scripture and science evidence.
- Young earth creationists filter all science facts through a Bible lens. If there is a conflict between science and a literal reading of Genesis, the science interpretation is always incorrect.
- Like young earth creationists, old earth creationists believe in special creation of all species but accept evidence for an ancient earth and universe.
- Intelligent design is not a view separate from creationism but is instead a way of defending such views. Intelligent design advocates accept evidence for an ancient universe but do not believe that evolution can explain the complexity of life.
- *Theistic evolution* and *evolutionary creationism* are two terms often applied to those who accept the evidence

for evolution and the age of the earth and who see no conflict between the evidence and religious faith.

- *Naturalism* and *scientism* are two terms often applied to those who view the natural, physical world to be all that exists. In this worldview, nothing is supernatural.

Discussion

1. Summarize each "camp." What are the primary beliefs of each?

2. Where are you camping? Do you find yourself with a foot in more than one camp?

3. In your faith journey, have you moved campsites? If you've moved, what was your motivation? Did you bring any artifacts from a previous camp with you?

4. In the big picture of faith, does it matter where you are camping? When might it matter? When might it not?

5. Consider the revealed character of God in Jesus Christ and in Scripture. How does each camp portray the character of God?

6. A popular chant taught widely in evangelical children's Bible classes says, "Absolutely true, absolutely true, everything the Bible says is absolutely true." How might this chant be perceived by children? By adults?

7. Consider this statement: the Bible can be trustworthy but not *literally* true regarding events of history and science. Do you agree or disagree? Explain.

8. Does unfettered curiosity lead to doubt? Is questioning ever a sin?

9. Why do you think some highly educated people in science (think Georgia Purdom, Nathaniel Jeanson, and Todd Wood) reject evolution?

10. Compare theistic evolution/evolutionary creationism to naturalism/scientism. Do these positions overlap? If so, where do they overlap? Where do they diverge?

Digging Deeper:
The Man Who Put Creation on the Calendar

He was a serious scholar and a clergyman—an archbishop in the Church of Ireland—and biblical history was his passion. He was a Protestant from a Catholic country, and he was determined to elevate Protestant scholarship beyond that of the Catholic Jesuits.

In the seventeenth century Archbishop James Ussher set out to locate his time and place within the larger context of biblical history. In doing so, he became an accidental expert in geology and a darling of young earth creationists.

Ussher saw his work as the continuation of the work of previous Jewish rabbis and early Christian scholars as he investigated key events in biblical history. When did Abraham first hear God's call? When was King David on the throne?[1]

Ussher's scholarship pinpointed many dates for biblical events, but only one date brought him centuries of fame: October 23, 4004 BC. More specifically, the evening *before*.

In his *Annals of the Old Testament*, Ussher declared October 23 to be the Date That Started It All: the creation of the universe.

Using the highest quality scholarship available at the time, Ussher triangulated data from ancient Hebrew and Egyptian texts; Chaldean, Persian, Greek, and Roman

histories; and biblical genealogies, kingdoms, and inter-
testamental periods. Ussher's creation date was compa-
rable to dates proposed by other notables: the Venera-
ble Bede, physicist Isaac Newton, astronomer Johannes
Kepler, and Cambridge scholar John Lightfoot.

Absent from Ussher's calculation of the creation date
was data from fossils, rock strata, or any other geological
observation. In fact, it would be more than a century before geologists and paleontologists proposed an ancient
earth. Not until the nineteenth century was a "hundreds
of millions of years" age established.

Bishop Ussher was not interested in what rocks or science said about creation. Instead, Bishop Ussher wanted
to establish God's providence in the timeline of human
history.

After Ussher's death a London bookseller began producing Bibles with Ussher's creation date printed in the
margins. The Bible was quite popular and widely sold,
but in addition to Ussher's calculations, it also included
illustrations considered a bit scandalous for the time.

After 1701 all editions of the King James Bible included "4004 BC" in the margin at the beginning of the
book of Genesis. Ussher's creation date took on the aura
of holy writ.

The *Scofield Reference Bible*, first published in 1909,
set the standard for study Bibles. It linked Scripture
together with a system of cross-referencing. It noted
common themes across books and chapters. It dated
the Bible chronologically, setting approximate dates for
events throughout the testaments. And Bishop Ussher's
date for creation, 4004 BC, was the date assigned for
Genesis chapter one.

Important to note is a comment added by C. I. Scofield himself: "the heaven and earth (in verse one) refers

to the dateless past," followed by a "catastrophic judgement which overthrew the primitive order."[2] Although Scofield allowed for an unknown "gap" of time between Genesis 1:1 and Genesis 1:2, the date 4004 BC was wholeheartedly adopted by Christian fundamentalists.

With the publication of Whitcomb and Morris's *The Genesis Flood*, the deal was sealed.[3] Credited with launching the modern creationist movement, *The Genesis Flood* endorsed a literal seven-day creation week, a literal worldwide flood, and dinosaurs living alongside humans. Whitcomb and Morris dated the earth at between five thousand and seven thousand years old—essentially Ussher's date, but reached for very different reasons.[4]

Currently the two leading creationist organizations, Answers in Genesis (of the Creation Museum and the Ark Encounter fame) and the Institute for Creation Research, cite Ussher's 4004 BC date in support of a young earth.[5] Despite all evidence to the contrary, Ussher's date supersedes geologic evidence.

And despite Ussher's theological intent, his date became geologic fact.

Resources

Kathryn Applegate and J. B. Stump, eds., *How I Changed My Mind About Evolution: Evangelicals Reflect on Faith and Science* (Downers Grove: InterVarsity Press, 2016).

> A collection of twenty-five firsthand stories from scientists, pastors, biblical scholars, and theologians about their journeys to acceptance of evolution. For some, it was the overwhelming evidence. For others, it was the intellectual dishonesty needed to "explain it all away." Many hid their acceptance of evolution from their faith community.

Karl W. Giberson and Francis Collins, *The Language of Science and Faith: Straight Answers to Genuine Questions* (Downers Grove: IVP Books, 2011).

> Each chapter in this excellent resource stands alone. The authors address nine important issues, one per chapter, including belief in evolution, science and the existence of God, and the fine tuning of the universe.

Ronald L. Numbers, *The Creationists: From Scientific Creationism to Intelligent Design* (Cambridge: Harvard University Press, 1992).

> At 606 pages, this book is a weighty read; it is a classic and outstanding resource. I use it as an encyclopedia for people and events in the context of the creationism-evolution discussion.

5

There Might Be a Time Machine in Your House

Chapter Highlights

- Young earth creationists reject an ancient universe and earth due to literal interpretations of biblical genealogies. Young earth creationists date the earth and universe at about six to ten thousand years old.
- The big bang theory states that the universe began as a tiny, incredibly hot, and incredibly dense single point, called a singularity.
- We can measure the distance from the earth to a star, and we know the speed of light. We can therefore know how long it takes light from a star to reach earth. All calculations point to a universe that is almost fourteen billion years old.
- We have eight hundred thousand years of sediment layers from lake beds and ice cores in addition to tree rings, indicating an earth far older than six thousand years.
- Radioactive decay in rocks points to an earth that is billions of years old.
- Evidence from the flipping of the earth's magnetic poles indicates an earth that is billions of years old.
- Evolution requires deep time. Rejection of evolution demands a young earth.

Discussion

1. What is the cosmic microwave background? What is its relationship to the big bang?

2. What evidence do we have for an expanding universe?

3. A six-thousand-year-old earth is not what it appears to be. Some argue that the earth was created full grown, like Adam, with the appearance of age. Is this view consistent with your view of Scripture? Your view of God? Why or why not?

4. Why is deep time necessary for evolution?

5. What evidence for an ancient universe is most convincing? What evidence is hardest to understand or accept?

6. If there was valid evidence for a young earth and universe, you would think that at least a few nonreligious scientists would support it. But that is not the case—advocates for a young earth/universe are always religious. Why might that be so?

7. Methods used to measure the age of the earth and universe are not unique to establishing age. Can a method be valid in one case but invalid in another? Explain.

8. Whitcomb and Morris, in their creationist classic *The Genesis Flood*, argue that both "parent" elements and the "daughter" elements (the result of radioactive decay) in radioactive rocks were created at the same time, probably in equilibrium. Do you trust the record of nature regarding radioactivity?

9. Instead of postulating that rocks or stars were created "full grown," some young earth creationists propose that the fundamental constants of nature have changed. For example, Answers in Genesis suggests that light has slowed down exponentially since the time of creation. All scientific evidence, however, points to the laws of

nature being consistent. Do you trust the record of nature regarding fundamental laws?

10.   If, as young earth creationists suggest, the earth was created with the appearance of age, why does the "apparent" age match so exactly with the scientific evidence? Some might say it looks rigged. Is this consistent with your conception of God?

Digging Deeper: Do You Trust the Science?

She was a working mom in a man's world when women could not yet vote. She won two Nobel Prizes—the first she shared with her husband, and the second she won on her own.

Marie Curie's research in radioactivity begot modern nuclear medicine.

When World War I broke out in France, Marie and her teenaged daughter outfitted a little car with a mobile X-ray machine, powered by the car's engine. Curie and her daughter drove the little car into the fight, revolutionizing the treatment of battlefield wounds. Eventually, Marie was able to outfit twenty such cars, known as "little Curies." More than a million soldiers were x-rayed in the field, and many were saved.

Radium, the radioactive element discovered by Marie and her husband, Pierre, was known to destroy abnormal skin growths. Doctors were understandably optimistic and believed radioactive elements could be used to fight cancer. And indeed, by the 1940s, radioactivity was used to successfully combat thyroid cancer. But although radioactivity could be used to treat cancer, we would later discover that it could also cause cancer.

At the end of her life Marie Curie suffered a variety of

ailments, probably due to her work with radioactive elements. She was only sixty-six when she died of leukemia, almost certainly brought on by her work with radium.

Modern nuclear medicine is much safer than it was in the early days. We use radioactive elements to both diagnose and treat patients. Modern practices maximize the benefit while keeping radioactive exposure to the patient at a minimum.

One key to safety in nuclear medicine is understanding the "half-lives" of radioactive elements; the time it takes for half of a radioactive element to decay into a stable element is called the "half-life." Some elements have extremely long half-lives—millions or billions of years—while others are as short as the blink of an eye. The shorter the half-life of the element, the less time a patient is exposed to radiation.

A nuclear stress test checks for blockages in arteries feeding the heart muscle. In the past, Thallium-201 was used to image a resting and then a stressed heart. But using this element was problematic. The images were often poor, and Thallium-201 has an alarming half-life of seventy-two hours. Technetium-99m, with a half-life of six hours, is currently used in stress tests. Not only are the images much better, but patient exposure to the radioactive element is greatly reduced.

Sometimes in nuclear medicine we need a longer half-life. We can mix Indium-111 with white blood cells and then track an infection in a patient's body. Indium-111's half-life of two and a half days allows it to stay active long enough to reach a remote infection.

The methods we trust for measuring half-lives of elements used in medicine are the same methods we use to measure half-lives of elements when we date rocks and determine the age of the earth. It's the same physics, the same chemistry, the same science.

And there's more.

The Fairbanks International Airport in Alaska recently renamed one of its runways. But don't get too attached to the new name—they'll probably have to rename it again in 2033.

Aviation is dependent on magnetic compasses for navigation: landing systems, air traffic control, and runway designations. Runways are identified according to the points on a compass, reflecting the compass reading to the nearest ten degrees. But once in a while, the compass flips. Earth's magnetic north and south poles swap locations.

If up is up, and down is down, what gives?

The molten iron inside the earth's core swishes, sloshes, and constantly moves. With enough flipping of individual iron atoms, magnetic north and south poles swap locations.

Magnetic pole flips are not uncommon. Poles have reversed 183 times in the last 83 million years. The intervals between flips vary, but the average is about 300,000 years. The last flip took place about 780,000 years ago.

Over the last few decades the north magnetic pole has been drifting toward Siberia.[1] Navigation safety requires we recognize the science behind the earth's magnetic field and adjust accordingly.

The methods we trust for measuring magnetic field shifts for aviation safety are the same methods we use to measure magnetic field shifts in determining the age of the earth.

It's the same geology, the same physics, the same science.

Resources

"EarthViewer," HHMI BioInteractive, October 16, 2017, https://
media.hhmi.org/biointeractive/earthviewer_web/earth
viewer.html.

> An interactive look at the earth throughout geologic
> history. Click "launch interactive" and explore different
> time periods. Rotate the earth to see different views.
> You can also overlay modern coastlines, cities, and geo-
> logical events by clicking on the "view" tab. All sorts
> of cool information about the earth through time.

"Learning about How the Universe Was Born: The Story of
the Horn Antenna," Nokia Bell Labs, August 6, 2018,
YouTube video, 3:47, https://youtu.be/lu2Go1ehx68.

> This short video about Penzias and Wilson's discovery
> of the cosmic microwave background, the leftover ra-
> diation from the big bang, features a short interview
> with Robert Wilson.

"Bristlecone Pines," Atlas Obscura, July 11, 2016, YouTube video,
3:56, https://youtu.be/Fv3u-atw9S8.

> An introduction to some of the oldest living organisms
> on earth—the bristlecone pine forest in Great Basin
> National Park in eastern Nevada.

6

It's Raining, It's Pouring, the Canyon Is Forming: Noah's Flood Explains It All

Chapter Highlights

- We find catastrophic flood stories in the collective memories of ancient cultures in the Middle East, Israel included.
- In the early twentieth century, "flood geology" proponents attributed earth's geologic features to Noah's global flood.
- In 1961 *The Genesis Flood* was published, expanding the idea of flood geology. The book is credited with establishing the modern young earth creationism movement.
- The widely released film *Is Genesis History?* features modern-day flood geologists and emphasizes the formation of the Grand Canyon by a worldwide catastrophic flood.
- Deposition, layering, rock formations, and other preserved artifacts in the Grand Canyon indicate a millions-of-years-old canyon formed over vast periods of time.
- Outside the canyon and across the planet, we find evidence of geologic features carved out over eons of time.
- A worldwide catastrophic flood is important to support Bible inerrancy, but important primarily as a support for a young earth. Flood geology is needed to explain the fossil record outside the context of evolution.

Discussion

1. Prior to reading this chapter, were you aware of non-Genesis flood and ark stories? What was your response when you first learned of these other flood stories?

2. Do ancient Middle Eastern context and culture inform our understanding of Genesis? How would you rate the influence of that culture and context on a scale of "interesting but not relevant" to "critical to our understanding"?

3. Review the history of "flood geology." What motivated the development of this concept?

4. What is "fining upward"? Do we see this phenomenon in the Grand Canyon?

5. We can observe in real time how deserts form, how salt deposits form, and the aftermath of raging floodwaters. If we know something from reality, how should we approach a religious text that contradicts it? Can the geologic record be trusted?

6. "Flood geologists" attribute many geologic phenomena to Noah's flood: shifting continents, rising mountains, and deep ocean vents. However, there are no references to these events or phenomena in the Bible. Can flood geology be defended as a biblical model?

7. If there were valid evidence for a worldwide catastrophic flood in geologic history, you would think that at least a few nonreligious scientists would support it. But like young earth advocates, proponents for a worldwide flood are always religious. Why might this be so?

8. For creationists, a worldwide flood is not as important scientifically as it is theologically. Why might that be so?

9. *Is Genesis History?* host Del Tackett puts his own spin on Dobzhansky's famous quote: "Nothing in the world makes sense except in light of Genesis."[1] What do you

think of Tackett's conclusion? Is such a framework workable or limiting?

10. Have you visited the Grand Canyon? If so, what were your impressions? Do you recall an emotional response to the enormity and beauty of the canyon?

Digging Deeper: Noah's Not the Only One

Genesis is not the only place we find a story of a great flood, complete with a boat builder, animals, and a bird sent to find dry land. There are multiple Middle Eastern flood stories remarkably parallel to the Noah story. And interestingly, they predate Genesis.

Atrahasis is one such Mesopotamian flood story. We have copies of *Atrahasis* that date as far back as the seventeenth century BC, but we don't know how old the story actually is.

In *Atrahasis* the god Enlil is annoyed with humans because they are just too noisy. Enlil decrees a flood to destroy the humans and restore some peace and quiet. The god Ea, however, intervenes and instructs Atrahasis to build a boat to save humanity. Atrahasis builds a boat, boards with his family, then seals the door with pitch.

British Museum Assyriologist Irving Finkel translated a recently discovered cuneiform tablet with sixty lines from the *Atrahasis* flood story. The boat in this story is round, made of rope, and animals board two by two.[2]

The *Epic of Gilgamesh* also includes a flood story. Our copies of *Gilgamesh* date at least to 2000 BC, but some scholars argue for a third-millennium date. The story is likely much older.

The flood story in *Gilgamesh* is strikingly similar to the Genesis story:[3]

- The boat builder is given specific dimensions for the boat.
- Animals are specified and loaded.
- The family boards.
- After the flood, the boat rests on a mountain.
- The boat builder releases a dove, then a swallow, but neither finds dry land.
- Finally, the boat builder releases a raven, but the raven does not return.

There is a Sumerian flood story even older than *Atrahasis* and *Gilgamesh*.

Israel grew up breathing the same cultural air as the Babylonians and the Sumerians. And of course there's the exile of Judah to Babylon. The Genesis flood story carries a different theological meaning than the earlier flood stories, but we can't ignore the resemblances.[4]

In 1998 Columbia University marine geologists William Ryan and Walter Pitman published research providing an interesting piece of the puzzle. They found evidence of a catastrophic localized flood that occurred in the Middle East about 7,500 years ago.[5]

The clues were fascinating.

The global climate was warming following the last ice age. Glaciers were melting and sea levels were rising.

Analyzing sediment cores, Ryan and Pitman discovered that the Black Sea was a freshwater lake in 5600 BC, not the saltwater sea it is today. Cores from the bottom of the Black Sea were from once-dry land. The lake was ringed with fertile farmlands and would have provided early farmers access to fresh shellfish.

The sediment cores were covered with uniform layers of mud, indications of a catastrophic flood.

Ryan and Pitman hypothesized that as sea levels rose,

the Mediterranean overflowed. Water rushed through the Bosporus Strait and into the Black Sea with a force many times greater than Niagara Falls. As water levels rose—maybe as much as six inches per day—human settlements were devastated. The deluge of salt water ruined the fertile soil.

Ryan and Pitman's hypothesis is supported by evidence from a National Geographic Society expedition. Led by Robert Ballard (best known for finding the *Titanic*), the expedition found well-preserved evidence of a stone-age human habitation and tools more than three hundred feet below the surface of the Black Sea.[6] Ballard's team also found fossils of freshwater species.

This ancient Middle Eastern flood was localized, not a worldwide event. There is no way to know if this flood was the inspiration for Middle Eastern flood stories, including Genesis. But it is plausible that such a devastating event would leave its mark on the collective memories of the people from the region, Israel included.

Assyriologist Irving Finkel writes, "There must have been a heritage memory of the destructive power of flood water, based on various terrible floods. And the people who survived would have been people in boats. You can imagine someone sunbathing in a canoe, half asleep, and waking up however long later and they're in the middle of the Persian Gulf, and that's the beginning of the flood story."[7]

Resources

Carol Hill, Gregg Davidson, Tim Helble, and Wayne Ranney, eds., *The Grand Canyon: Monument to an Ancient Earth* (Grand Rapids: Kregel Publications, 2016).

A beautifully illustrated resource book with user-friendly explanations of the geology needed to understand the formation of the Grand Canyon. The contributing authors argue the canyon is ancient and not the result of a worldwide flood.

Irving Finkel, "The Ark Before Noah: A Great Adventure," Oriental Institute, July 20, 2016, YouTube video, 58:18, https://youtu.be/s_fkpZSnz2I.

A delightful lecture by Irving Finkel, translator of the "round ark" tablet in the British Museum. Finkel is a funny and informative storyteller. Finkel's translation of the cuneiform tablet necessitated a whole new look at the Noah story.

"Grand Canyon National Park," National Park Service, March 12, 2022, https://www.nps.gov/grca/index.htm.

This website has links to information about the ecology, geology, and human history of the Grand Canyon. It also includes virtual tours and photo galleries.

Is Genesis History?, https://isgenesishistory.com.

A film narrated by young earth creationist Del Tackett. It includes interviews with multiple creationists who support a young earth and a Grand Canyon created by Noah's flood. The film presents "flood geology" by adherents to the position. It is available for purchase and on streaming sites.

7

The Flood and the Fossil Record

Chapter Highlights

- Evidence from geology and paleontology confounded the Victorian world. The abundance of fossils discovered and identified in the nineteenth century raised new theological questions.
- Mary Anning, a Victorian era fossilist, was responsible for many magnificent finds of the era.
- Both young and old earth creationists acknowledge that the fossil record challenges the Genesis account. However, any conflict between science and the Bible is always decided in favor of the Bible.
- Hydraulic sorting is a creationist explanation for fossil layers. According to this explanation, organisms destroyed in Noah's flood are found sorted by weight, mobility, and intelligence.
- There is a recognizable order in the fossil record. A catastrophic flood could not produce the predictable order we find.
- It is difficult to ascribe the "hydraulic sorting" explanation to plant fossils. Plant fossils are rarely addressed by creationists.
- The unique collection of mammals found in Australia challenges the flood dispersal narrative.

Discussion

1. What role did Mary Anning play in the fledgling science of paleontology?

2. Describe the patterns we find in the fossil record. What do we *not* find?

3. Many creationists use a Bible lens when interpreting scientific evidence. In other words, when scientific evidence contradicts a literal reading of Scripture, the Bible always wins out. What are possible consequences of using this approach to evidence?

4. Describe the concept of hydraulic sorting. Can you make a case for both hydraulic sorting and evolution?

5. In the fossil record, plants that bear seeds and fruit are relative newcomers. What does Genesis say about the creation of seed- and fruit-bearing plants?

6. Have you experienced a devastating flood like the flood in New Orleans following Hurricane Katrina? Describe the experience. If you haven't, describe photos and videos you've seen of devastating floods following storms or tsunamis. What happens to plant life? What happens to animals caught in the flood? What about human victims? How would you describe the aftermath of a catastrophic flood?

7. What makes a documentary, news story, or historical document "true" by twenty-first-century standards? What are your expectations regarding dates and times? What are your expectations regarding sequence of events? What are your expectations regarding names and quotations? Are these realistic expectations of a Bronze Age or Iron Age text? Why or why not?

8. In the fossil record, millions of years pass between the last appearance of dinosaurs and the first appearance

of humans. Many creationists believe humans and dinosaurs coexisted. Biblical references to "leviathan" and "behemoth" are interpreted as human descriptions of the dinosaurs that lived among them. Ancient artwork with dragon motifs is also said to depict dinosaurs. Do you find these speculations plausible? Explain.

9. Do you find a catastrophic worldwide flood a credible explanation for the fossil record? If so, what would change your mind? If not, what would change your mind?

10. Are there "two sides" regarding the fossil record? Is it possible to have two legitimate interpretations of the record?

Digging Deeper:
The Man Who Almost Scooped Darwin

He was a nineteenth-century equivalent of a modern-day twenty-something backpacker—broke, sleeping under the stars, living on crackers. Meet Alfred Wallace.

The year is 1848. Wallace and his buddy set off to explore Brazil. Both are fascinated by the beetles, butterflies, and birds found in the Amazon. Although Wallace received a good education back in England, his family does not have money. Wallace arranges to sell the best of his finds to museums and collectors. Wallace is a scientist at heart but collecting is a means to support himself.

Wallace and his buddy split up to explore different areas of Brazil. Traveling alone or with locals, Wallace is in heaven. The collections he gathers are amazing: exotic butterflies, spider monkeys, extravagant birds. And his notes. Copious notes that record where the various species live and, importantly, contain Wallace's ponderings: where do species come from?

Four years in the jungle. Four years of collecting and annotating.

On the voyage back to England, the ship catches fire and sinks. Wallace escapes with his life, but little else. He manages to save one small box of notes and sketches. After ten days adrift in a leaky lifeboat with inadequate food and water, the survivors are rescued. Wallace vows never to sail again.

A few years later Wallace meets Charles Darwin. Darwin is already famous—the most famous naturalist in England. Darwin's published accounts of his trip around the world on the *Beagle* are popular.

As Wallace enthusiastically shares his ponderings about the origin of species, Darwin is taken aback. Darwin had been privately writing his thoughts about the origin of species for years but had shared them with only a few close friends and colleagues.

Darwin is a man of standing. He is married and has a family to support. And to suggest anything other than an instantaneous, special creation of all species is, well, dangerous.

Darwin isn't ready.

Driven by his questions about the origin of species, Wallace heads out on a new voyage to the Malay Archipelago. For eight years, Wallace hops island to island, collecting, writing, thinking. He observes. He collects. Wallace amasses over one hundred thousand insects and thousands of shells, bird skins, mammals, and reptiles—many new to science.

But the questions just won't stop: Why are butterflies in Malaysia so different from butterflies in the Amazon? Why do cockatoos live in the Archipelago, but macaws and hummingbirds live in the Americas? Why do species arise near similar species? Are species connected to each other?

And here's a puzzle: manatees are mammals who spend their entire lives in water. Using flippers, manatees swim and dive and make their home with the fishes. But look inside the flipper of a manatee, or a whale for that matter, and what do you find?

Fingers. Useless fingers locked inside a flipper.

What?

Wallace can't escape the thought that species arise from other species, modified from a previous form.

And here's another mystery: there seems to be an invisible line separating the western and eastern islands of the Archipelago. Although the islands are very close geographically and very similar ecologically, the animals living on opposite sides of the line are very different.

The animals of the western islands are like those of Asia—tigers, monkeys, squirrels, and orangutans—all placental mammals. Animals of the eastern islands are like those found in Australia—kangaroos and other pouched marsupials and those weirdo egg-laying mammals, the monotremes. No marsupials or monotremes in the west; only a few placentals that flew or swam there in the east.

It must have been, mused Wallace, that the eastern islands were once connected to Australia, and the western islands were once part of Asia.

Wallace was onto something.

Special creation could not explain the invisible line, but Wallace could: species arise from other species. And if a species is geographically isolated, the species will evolve independently in their own region.

(Wallace was correct. In the geologic past, sea levels were lower in the area and the western islands were part of Asia. Likewise, the eastern islands were attached to Australia. Today this invisible line is recognized as the "Wallace Line.")[1]

At first Wallace does not understand the "why" of what he hypothesized. But it eventually comes to him while he is sick in bed with malaria.

Wracked with fever, Wallace nonetheless writes his old pal and fellow naturalist Charles Darwin. Wallace thinks he knows the answers to his burning questions. Species arise from previous species. And species change because the fittest individuals adapt to their environments and leave more offspring.

Reading the letter, Darwin knows his old friend had discovered what he himself had known for years but was too intimidated to publish. With Wallace's blessing, both Darwin's and Wallace's papers on evolution by natural selection are read aloud to the science community in London.

But it was Darwin who published first, his classic *On the Origin of Species*, and it is Darwin who is most closely tied to the theory of evolution.

It is hard in the twenty-first century to grasp the enormity of Darwin's and Wallace's work. At a time when almost everyone in their world thought species instantaneously appeared, specially and separately made, Darwin and Wallace and many other Victorian paleontologists gave evidence to the contrary.

The conclusion was irrefutable.

Resources

We Believe in Dinosaurs, https://www.webelieveindinosaurs .net.

A fascinating look at the "Ark Encounter" in Kentucky, this film is available for streaming on demand from multiple sources. The ark, built to be life-size, intends to prove that the Bible is scientifically and historically accurate, the earth is young, and evolution is a lie. The film is a behind-the-scenes look with the supervisor of art and design—the man responsible for creating the animals for the ark, including the baby dinosaurs. There are extensive interviews with a local geologist who has long fought against the non-science perpetrated by the ark and the Creation Museum. David MacMillan, a former creationist and former lifetime member of the Ark Encounter (his name is on a plaque and everything!), is featured.

Lauren Anholt, *Stone Girl, Bone Girl* (London: Frances Lincoln Children's Books Publishing, 2006).

Catherine Brighton, *The Fossil Girl* (London: Frances Lincoln Children's Books Publishing, 2007).

I love quality children's books for an introduction to a topic. Here are two of my favorites:

Stone Girl, Bone Girl is the story of Mary Anning's early discoveries told in the context of her family, her community, and the rising popularity of Victorian fossil collecting.

The Fossil Girl is a colorful picture book in the format of a fun graphic novel. An afterword page describes the importance of Mary's finds to Charles Darwin's work as well as the angst her finds triggered in some religious people.

Patricia Pierce, *Jurassic Mary: Mary Anning and the Primeval Monsters* (Stroud, Gloucestershire, UK: The History Press Publishing, 2006).

A wonderful book for adults and older kids, with sketches by Mary (including one she made of her little dog, Tray), photos and drawings of Lyme Regis, and interesting science asides throughout the chapters.

James McNish, "Who was Alfred Russel Wallace?," Natural History Museum, n.d., https://www.nhm.ac.uk/discover /who-was-alfred-russel-wallace.html.

From the Collections of the Natural History Museum, London. Explore gorgeous drawings and specimens from the Alfred Russel Wallace collections. Links to Wallace's correspondence and extensive photos of his bird collection.

"Bill Bailey on Alfred Russel Wallace," Natural History Museum, YouTube video, 6:34, July 3, 2013, https://youtu .be/KT2YbugYcjQ.

Comedian Bill Bailey, a huge fan of Alfred Wallace, visits the Natural History Museum in London. This video provides a short introduction to a field of science founded by Wallace—the study of the geographical distribution of animals.

8

Written in Stone

Chapter Highlights

- The fossil record is an unfolding, sequential panoramic view of the development of life on earth. Trends in body architecture can be identified in the fossil record.
- The fossil record is biased against soft bodies. It is biased in favor of hard body parts like shells, bones, and teeth; it is biased in favor of common and long-lived species.
- The earliest life forms in the fossil record are mats of single-celled bacteria. Early multicellular life includes soft-bodied fern- or feather-shaped animals.
- In the layers of the Cambrian, we find tremendous diversity—primitive versions of almost all modern body plans.
- Despite the diversity and relative "suddenness" (eighty million years) of the Cambrian, we find no mammals, fish, dinosaurs, humans, or any kind of vertebrate.
- Amphibians, the first four-limbed animals on land, grew large and were the apex predators of their day.
- The amniotic egg (an egg that does not dry out on land) was a landmark in the evolution of four-limbed animals. The evolution of the amniotic egg paved the way for dispersal of tetrapods away from water's edge.

- The end of the age of giant reptiles saw the rise of mammals, and eventually humans.

Discussion

1. William Smith coined the term "faunal succession." What was Smith describing? What led him to this conclusion?

2. Why is the fossil record described as panoramic "snapshots" of life and not an exhaustive history of life?

3. Describe the various ways fossils are formed. What kinds of things most readily fossilize?

4. The panoramic picture of life found in the fossil record is incomplete—we do not have a record of every species. We cannot identify direct ancestors with certainty. Of what value is the fossil record?

5. My husband has a childhood memory of a picture book at his grandmother's house. It was all about dinosaurs— and how they were "fake." It was a religiously motivated interpretation of the fossil record. Have you ever read (or been told) that fossils are "fake"? What was the rationale for such an explanation?

6. Along with out-and-out fakery, other explanations for the fossil record run the gamut from "scientists are hiding the truth" to "remains of animals that died before (or during) the flood." Others simply ignore the question. What are your thoughts regarding fossils?

7. There is a small town outside of Fort Worth, Texas, that is a mecca for both scientists and young earth creationists alike. Glen Rose, home of Dinosaur Valley State Park, boasts some spectacular dinosaur footprints. According to many creationists, Glen Rose is also home to human prints alongside the dinosaur prints. The human "prints" have long been discredited, but that does not deter the local

creationism museum and other creationists in the area. "Most everyone in Glen Rose that I know believes man and dinosaurs coexisted," says one local woman. "The only conflict we have is when people move from metropolitan areas and have different value systems."[1] What do you think? Are there cultural/social/economic/urban/rural divides in the creation-evolution discussion? If so, which divides are most prevalent? Why do you think this is so?

8. Here's another comment from a Glen Rose local: "I'm religious . . . and I know God made it all, but I don't know or care if he made it in billions of years, or if he put time zones in there to make it look like billions of years."[2] Is this a reasonable conclusion? Is this an intellectually satisfying conclusion?

9. The fossil record and the Genesis creation story conflict in timeframe and order. Of what value is the creation story to someone who accepts the fossil record?

10. When asked what evidence would disprove evolution, J. B. S. Haldane famously answered, "Fossil rabbits in the Precambrian." Why would Precambrian rabbits disprove evolution?

Digging Deeper: Life, Death, and Life After Death

It was six miles wide—the size of Mount Everest. It weighed hundreds of billions of tons. It was traveling eighty thousand kilometers per hour, hurtling toward earth at a rate twenty times faster than a speeding bullet. At that speed, friction would heat the surrounding air to a temperature several times hotter than the sun. On impact, the energy equaled one hundred million nuclear bombs exploding all at once. Within one thousand kilometers of the impact, all life broiled. Literally.

Sixty-six million years ago an asteroid rained extra-terrestrial death and disaster on the earth. There have been five great extinctions in the history of the earth—this the most recent. Sean B. Carroll calls it "the day the Mesozoic died."[3]

And life was never the same.

The Yucatan Peninsula in Mexico was ground zero. Debris, soot, and vaporized rock blasted into the atmosphere. Some of it orbited the earth, then rained back down. Much of it stayed in the atmosphere, blocking out the sun. Wildfires raged, adding smoke to the air. No sunlight, no photosynthesis.

Plants died. Plant eaters died. Animals that eat plant eaters died. Food chains collapsed. Sixty to eighty percent of bird and amphibian species went extinct. The mighty dinosaurs, dominators of the planet for 180 million years—all gone. Along with the dinosaurs went enormous flying reptiles, the pterosaurs, and monstrous predatory reptiles in the ocean, the mosasaurs. In some parts of the world, more than half of plant species disappeared.

Survivors were small and lived in the water or burrowed underground. Small animals also died in the asteroid aftermath, but enough survived to carry on. Winners included a lone branch of the dinosaur tree—birds—and tiny mouse-like mammals.

For years the event that triggered this mass extinction was a mystery.

A curious clue was found in a seam of clay running through rock layers of the same age across the globe. Called the K-T boundary, the line delineates in stone the world of the dinosaurs and the world after dinosaurs. Below the line, dinosaur fossils. Above, none.

The clay in the K-T boundary is rich in iridium, an element rarely found in the earth's crust. A tiny amount of iridium constantly rains down on earth from space— just a light dusting. The amount of iridium in the K-T boundary was thirty times greater than that in the surrounding rocks, far more than would be expected from a light space dusting. The K-T boundary sites around the world had similarly high levels of iridium.

Iridium is rare on earth, but guess where we find high levels of iridium? Meteors and asteroids.

In the late 1970s Nobel Prize–winning physicist Luis Alvarez proposed an answer. The source of the iridium in the K-T boundary was an enormous asteroid, vaporized on impact, that then rained down to earth.

The missing piece of the puzzle was a crater of the right age, type, and size. The search led to Mexico and an underwater crater first mapped in 1978. The Chicxulub crater in the Yucatan Peninsula met all three criteria.

By the early 1990s Alvarez and his team found the definitive clue. Nuclear bomb explosions damage the surrounding rock in a characteristic way. The shock of the impact dislocates minerals in the rock and produces "shocked quartz."

The Chicxulub crater is full of shocked quartz. Asteroid ground zero.

Eventually, the air cleared, and once again the sun shone. The world was on the road to recovery.

Near Denver, Colorado, are one million years of exposed rock, just above the K-T boundary. In these rocks we have a front-row seat to the recovery that led to our time, our world.[4]

What do the first post-dinosaur, post-asteroid fossils tell us?

Right above the K-T boundary are a plethora of fungal spores. Fungus grows on dead things, and most of the world was dead and rotting.

Then we begin to see ferns. Ferns are often the first plants to return after a forest fire and are signs that plant life is recovering. Palms are next, a wealth of palms. A forest of palms. By three hundred thousand years after the asteroid, plant life rebounds.

We also find animals: turtles, crocodiles, and mammals. During the dinosaur days, mammals were small; most were the size of a mouse. By three hundred thousand years post-asteroid we find mammals the size of beavers, and by seven hundred thousand years we find mammals weighing more than one hundred pounds.

In the absence of dinosaurs, the mammals were taking off.

At seven hundred thousand years into recovery, we find the first fossils of a legume. Legumes are packed with protein and likely fueled the increase in mammal size.

The recovered world was not to be ruled by giant reptiles. Instead, it became the domain of mammals—giants in the ocean and on land, hooved runners, fur-covered creatures filling a world of niches left vacant by dinosaurs. And of course, primates who eventually came down from the trees, walked on two legs, controlled fire, and made tools.

Resources

Donald R. Prothero, *The Story of Life in 25 Fossils* (New York: Columbia University Press, 2015).

   The story of evolution told in twenty-five landmark fossils. The book is a wealth of information about fossil hunting and the importance of each find. Chapters include a "See It For Yourself!" section listing museums where the featured specimens can be seen.

Paul D. Taylor and Aaron O'Dea, *A History of Life in 100 Fossils* (London: The Natural History Museum, 2014).

The full-color, high-resolution photographs in this beautiful "coffee table" book are organized by geologic time: the Precambrian, the Paleozoic, the Mesozoic, and the Cenozoic. This book is a trip through a paleontology museum.

Ashley Hall, *Fossils for Kids: A Junior Scientist's Guide to Dinosaur Bones, Ancient Animals, and Prehistoric Life on Earth* (Emeryville, CA: Rockridge Press, 2020).

This handy little book, a wonderful introduction for kids and their adults, contains photographs, charts, definitions, and descriptions.

"The Day the Mesozoic Died," HHMI BioInteractive, October 11, 2012, video, 33:43, https://www.biointeractive.org /classroom-resources/day-mesozoic-died.

Why did the dinosaurs disappear? It was a mystery until a team of scientists followed the evidence written in stone. Find links to associated resources and tips for using this short film.

"Out of the Ashes: Dawn of the Age of Mammals," HHMI Bio-Interactive, May 1, 2020, video, 16:11, https://www.bio interactive.org/classroom-resources/out-ashes-dawn -age-mammals.

After the last great extinction, how did life recover? A treasure trove of evidence was found in Colorado. This is a short film with links to related content and a film guide.

"How to Find a Dinosaur," HHMI BioInteractive, April 18, 2012, video, 4:05, https://www.biointeractive.org/classroom -resources/how-find-dinosaur.

Go into the field with a paleontologist and see how fossils are found.

"The Making of Mass Extinctions," Howard Hughes Medical Institute, February 3, 2022, https://media.hhmi.org/bio interactive/click/extinctions/.

This interactive resource looks at the five great extinctions in earth's history and allows you to locate each extinction in time and read a few details about each one.

9

In Search of the Missing Missing Link

Chapter Highlights

- The most misunderstood aspect of evolution is common ancestry.
- The fossil record is full of "missing links" or transitionals as biologists call them.
- *Tiktaalik* is key to understanding fish-to-land evolution.
- The dinosaur-to-modern-bird transition is one of the best documented in the fossil record.
- Understanding the transition of whales from land animals to sea creatures requires evidence from both modern genetics and the fossil record.
- Control genes direct body architecture. A single change in control genes can have a dramatic effect and result in new body forms.

Discussion

1. Before you read this chapter, how would you have defined a "missing link"? What images come to mind?
2. Do you accept the existence of transitionals, a.k.a. "missing links"? If not, what makes them seem impossible? If you accept the existence of transitionals, what evidence is most convincing?

3. How did the theory of evolution guide Neil Shubin in his search for a water-to-land transitional?

4. What can we learn about the evolution of feathers from the many feathered dinosaurs we have found?

5. Evolution is best represented by a tree or a bush, not a straight line. Why is this so?

6. The term "living fossil" is not a biological term but is often used to describe living species that have remained relatively unchanged for millions of years. Horseshoe crabs, coelacanths, crocodiles, and ginkgo trees are well-known "living fossils." Using what you know about evolution and natural selection, why might some species remain unchanged for millions of years?

7. Somewhere in your family tree, you probably have a great-aunt. You definitely have a great-grandmother. You are directly descended from your great-grandmother, but your great-aunt provides a reasonable picture of the lineage that led to you. A common misconception about how evolution works is that there is a direct line of descent from one primitive form to the next, until at last, the modern creature arrives on the scene. How does this misconception shape understanding about transitionals? How do transitionals inform us of the general trajectory of evolution?

8. Explain the orchestra analogy regarding control genes.

9. Control genes "conduct the orchestra" in our DNA. How might control genes explain the wide diversity of body forms we see in the Cambrian?

10. Were you surprised to learn that the number of genes in an organism's DNA does not correlate with complexity? Comparing total amount of DNA in species is also surprising. An amoeba has more than 600,000 base pairs, but the human genome contains only 3,400 base pairs.

What does this say about the place of humans in the broader context of living things?

Digging Deeper:
"What Sensitive Ears You Have, Granny!"

It was rather restrictive.

When the first fish descendants ventured out onto land, they filled their lungs with air and hoisted themselves along with skeletons capable of supporting them outside the water. I like to image they rejoiced (if possible, with their fishy brains) at their newfound freedom.

But the first tetrapods could not venture too far afield. Their fish-like, shell-less eggs would dry out unless laid in water.

The amniotic egg was a major landmark in evolution. An amniotic egg surrounds the developing embryo with fluid-filled membranes and a protective shell. A whole new world opened up. No longer tied to water's edge, the amniotes filled niches all over the planet.

The tremendous advantage afforded the amniotes led to great diversity. About 290 million years ago, the amniotes split into two major lineages. One lineage led to reptiles: dinosaurs, crocodiles, snakes, lizards, turtles, and birds. The other lineage led to the synapsids.

Synapsids are often called "mammal-like reptiles," but they were not reptiles, and they were not mammals. Synapsids were a branch of amniotes, a lineage that eventually gave rise to mammals.

From our vantage point, the synapsids were transitionals on the road to mammals. But in their day, they were simply the creatures they were—adapted to their environment and successful.

The fossil record provides an amazing picture of the evolution of mammals from the very reptile-like synapsids. In the synapsids, we see the earliest evidence of one of the defining characteristics of mammals: the unique mammalian ear.

Mammal ears are extraordinarily sensitive—more so than any other animal. The key to the superiority of the mammalian ear lies inside the skull: the three bones of the middle ear. Reptiles only have one bone in the middle ear. The extra two bones allow for more amplification of sounds from the eardrum.

The transition to the three-boned mammalian middle ear from a one-boned reptile ear began in the jaw.

The earliest synapsids, like their close reptile relatives, had a large lower jaw made of three bones: a large bone carrying the teeth and two smaller bones forming a hinge with the skull.

Over the next eighty-five million years in synapsid evolution, the large bone carrying the teeth trended larger and larger. The two smaller jaw bones trended smaller and smaller. In the very late synapsids and the earliest mammals, the bone carrying the teeth grew so large it made contact with the skull, forming a new joint.

The two smaller bones trended progressively tinier and tinier and eventually dislocated from the large bone. The two tiny bones migrated to the middle ear, joining the single bone of their reptile ancestors.

As mammal embryos develop in the womb, two bones migrate from the jaw to the ear. Marsupial mammals (pouched mammals like kangaroos and opossums) are born extremely early and underdeveloped and can show us embryonic development outside the womb.

Opossum babies are born so early, they are only a few millimeters long, with no hair and no hind limbs. Opossum babies give us a picture-perfect view of the transition from a reptilian to a mammalian-style ear.

Biologist Karen Sears euthanized short-tailed opossum babies in consecutive stages, from birth to adult.[1] Stripping away the flesh with maggots (yuck!), Sears created an incremental picture of ear development in the opossum.

At birth the opossum has a single ear bone and two extra bones in the lower jaw, like a reptile. As the opossum grows, the lower jaw gets progressively larger, and the two extra bones get progressively smaller. By adulthood, the two tiny extra bones have moved up into the ear.

Born with a "reptile trait," Sears's opossums help us understand a key transition in the evolution of mammals.

Resources

Your Inner Fish, episode 1, "Your Inner Fish," aired April 9, 2014, on PBS. https://www.pbs.org/your-inner-fish/watch/.
Your Inner Fish, episode 2, "Your Inner Reptile," aired April 16, 2014, on PBS. https://www.pbs.org/your-inner-fish/watch/.

> Two episodes in a three-part PBS series featuring Dr. Neil Shubin. Based on Shubin's best-selling book *Your Inner Fish*, all three episodes in the series are must-sees!
>
> The first episode explores the origin of four-limbed animals from fish ancestors. The film features *Tiktaalik*, Shubin's most famous fossil discovery. The second episode focuses on the origin of mammals, tracing our hair, skin, teeth, jaws, and keen sense of hearing back to our reptile ancestors.

"Great Transitions: The Origin of Tetrapods," HHMI BioInteractive, video, 17:11, March 28, 2014, https://www.bio
interactive.org/classroom-resources/great-transitions
-origin-tetrapods.

Narrated by Neil Shubin, this short film hits the highlights of the discovery of *Tiktaalik*. It is a great sum-
mary and a wonderful introduction to transitionals.

"Great Transitions: The Origin of Birds," HHMI BioInteractive,
video, 18:59, February 12, 2015, https://www.biointerac
tive.org/classroom-resources/great-transitions-origin
-birds.

The dinosaur-to-bird transition is one of the best-
documented transitions in the fossil record. From the
first discovery of *Archaeopteryx* to a treasure trove of
feathered dinosaurs in China, this short film is an ex-
cellent introduction to the origin of modern birds.

"Exploring Transitional Fossils," HHMI BioInteractive, May 30,
2014, https://www.biointeractive.org/classroom-re
sources/exploring-transitional-fossils/.

An easy interactive exploring several transitional fossils
in the evolution of tetrapods.

Dougal Dixon, *When the Whales Walked* (Lake Forest, CA:
Quarto Publishing, 2018).

A beautifully illustrated picture book featuring important landmarks in evolution. The book features whales,
but it also shows many other important transitions in
the fossil record.

10

It's All or Nothing: Intelligent Design

Chapter Highlights

- The intelligent design model has replaced traditional creationism for many religious people, but both camps of creationism (young and old) incorporate concepts of design.
- Intelligent design was born out of banning creationism in public schools.
- The centerpiece of intelligent design is the idea of irreducible complexity; life is so complex it could not have simply evolved.
- In *Kitzmiller vs. Dover*, a federal court ruled that intelligent design is inherently religious, and as such, it cannot be taught as science in public schools.
- Intelligent design accepts the fossil record, but species are designed, not descended, from a common ancestor.
- Broken genes in the human genome pose a problem for intelligent design.
- Partial pathways and repurposed structures refute claims of irreducible complexity.

Discussion

1. Why is intelligent design an attractive alternative to traditional creationism?

2.   Explain "irreducible complexity," the principle under-
     lying intelligent design. How is a mousetrap used to
     defend irreducible complexity? How does a mousetrap
     refute irreducible complexity?

3.   The intelligent design textbook *Of Pandas and People*
     was a reworking of a creationist textbook, substituting
     the term "designed" for "created." What does this imply
     about the intelligent design model?

4.   The human genome is littered with broken genes and
     leftovers from ancient viral infections. Do these facts
     suggest common ancestry? Might a designer mimic
     evolution? Why or why not?

5.   Intelligent design advocates often concede "microevo-
     lution" (small changes allowing for adaptation to the
     environment) but deny "macroevolution" (change re-
     sulting in new species). Although these terms are not
     used in biology, is this an intellectually satisfying dis-
     tinction for you? Why or why not?

6.   Back problems are common in humans. Evolution tells
     us that we began with a horizontal spine and stood it
     upright. In the process, weight balance is off, and an
     S-shaped curve evolved to compensate. The result is
     a back architecture fraught with problems and prone
     to fracture and rupture of discs. Intelligent design says
     that the human body was designed by a designer, un-
     derstood to be God. Can you defend the human back as
     the product of an intelligent designer? Explain.

7.   Intelligent design apologetics often focus on the un-
     explained in science. In other words, if we can't fully
     explain the source or the functioning of a structure or
     a process, it is assumed to be designed. What problems
     do you see with this line of thinking? What happens
     when something previously unknown is explained?

8. "Evolution is a tinkerer." What evidence supports this statement?

9. Many people are attracted to intelligent design because the alternative appears to be deism—God simply "wound the clock" on creation and then stepped out of the picture. Is intelligent design the alternative to deism? Does evolution require a deistic view of God? In your understanding, is it the nature of God to be a micromanager?

10. The idea of God's sovereignty was central to John Calvin's doctrine. Many modern Christian denominations are influenced by Calvin, some to a greater degree, others to a lesser degree. Regardless, a "God's got this" or "God's in charge" mentality pervades western Christianity. Do you see Calvin's influence in the attraction to intelligent design? If so, how?

Digging Deeper: Dandy Designs?

Which animal kills more people than any other animal on the planet? A fearsome lion? An unpredictable hippopotamus? A venomous snake? Sharks, right? After all, there is a whole week of shark television devoted to scaring us out of the water!

The answer is "none of the above."

The most dangerous animal on earth is . . . the mosquito.

Mosquitos carry all sorts of horrible diseases—dengue, yellow fever, encephalitis—but the worst by far is malaria. Malaria kills hundreds of thousands worldwide every year, mostly young children and pregnant women, and mostly in poor areas of the world.

British physician Anthony Allison grew up in the Rift Valley of Kenya. Allison is responsible for a historic breakthrough in our understanding of malaria, but malaria wasn't originally on his radar. Allison returned to Africa in the 1950s with the intent of studying the ABO blood groups in East African people, but he soon turned his attention to the sickle cell condition.

We carry the instructions to make hemoglobin, the oxygen-carrying component in blood, in our DNA. Just one single change, a mutation at a single point in the hemoglobin gene, results in a defective version of hemoglobin.

And what a difference one little point mutation makes. Healthy red blood cells have a characteristic biconcave shape—a smooth disk with a depressed center. This shape allows a red blood cell to squeeze itself into the tiniest of blood vessels to deliver its oxygen payload. Red blood cells with the defective form of hemoglobin are hard and sticky. Instead of smooth and biconcave, the cells are C-shaped, like a crescent or sickle.

Humans have two copies of every gene—one copy from mom, one copy from dad. If one copy of the hemoglobin gene is operational, enough functioning red blood cells are made to keep a person healthy. A carrier of a nonfunctioning hemoglobin gene may never know it.

It was the carriers who caught Dr. Allison's attention.

Allison noticed a peculiar pattern. In equatorial Africa, as many as forty percent of the population carried one copy of the mutated hemoglobin gene. But that's not all—carriers were clustered on the coast and near Lake Victoria.[1] In the highlands of Kenya, however, carrier frequency dropped dramatically.

Allison knew that the warm, wet areas of Kenya were breeding grounds for the *Anopheles* mosquito, carrier of the deadly malaria parasite. The pieces were coming together.

Natural selection preserves beneficial traits—traits that confer some sort of advantage in a particular environment. Is there a correlation between malaria and the defective hemoglobin gene? Allison wanted the answer.

In a massive study, Allison collected more than five thousand blood samples across East Africa. Soon it all fell into place: carriers of the defective gene are resistant to malaria.

Carriers of the defective gene make a few sickled cells, enough to make their blood inhospitable to the malaria parasite, but not enough to impact blood circulation. Allison's observation holds true not only in East Africa but wherever malaria is endemic.

To intelligent design advocates, the human genome bears the mark of design. Random mutations only break things. Important mutations, mutations that confer benefit, cannot arise randomly and therefore must have been directed, micromanaged by an intelligent designer.

Mosquitos kill, but a mutation in one gene confers malaria resistance. A good thing, right? A dandy design! With a tweak at just the right spot in the DNA, the designer heroically saves lives.

Not so fast.

In a population with a high frequency of a single gene, it won't be long before two carriers have children. If both parents are carriers, chances are one in four that a child will inherit not one, but two copies of the defective hemoglobin gene. Children who inherit two copies of the gene are resistant to malaria, yes, but at a terrific cost: sickle cell disease.

Red blood cells stick and clump and block blood vessels. The result is excruciating pain and often stroke. Sickled cells are short-lived, so sufferers are anemic, exhausted, and prone to serious infections and kidney failure. Before modern medicine, children with sickle cell disease usually died in early childhood. Many still do.

If people with two copies of the sickle cell gene are terribly ill or die in childhood, why the high frequency of the gene in malaria-endemic areas? The malaria resistance afforded to those with one copy of the gene fuels its spread.

Crediting the malaria-resistance mutation to a loving choice by an intelligent designer demands consideration of the other side of the coin: the three hundred thousand children born each year with sickle cell disease.

Resources

"Judgement Day: Intelligent Design on Trial," November 3, 2011, YouTube video, 1:51:25, https://youtu.be/x2xyrel-2vI.

> In 2005, the school board in Dover, Pennsylvania, ordered high school biology teachers to include intelligent design in their classrooms as an "alternative" to evolution. This is a fascinating NOVA documentary about the subsequent trial in federal court in which intelligent design was ruled to be a religious construct, not science.
>
> You can read about the documentary on the PBS website: "Judgement Day: Intelligent Design on Trial," PBS, 2007, https://www.pbs.org/wgbh/nova/video /judgment-day-intelligent-design-on-trial/.

Kenneth R. Miller, *Only a Theory: Evolution and the Battle for America's Soul* (New York: Viking, 2008).

> The Dover trial is to the twenty-first century what the Scopes trial was to the twentieth. Miller, a key witness at

the trial, examines the case that ultimately ruled intelligent design to be an inherently religious concept. Miller explores public perceptions of science and other cultural trends that underlie the appeal of intelligent design.

"The Wedge Document," National Center for Science Education, October 14, 2008, https://ncse.ngo/wedge-document.

Full text of the Wedge document provided by the National Center for Science Education. The Wedge document was created by Phillip Johnson as a fundraising tool for the Discovery Institute, an intelligent design think tank. Originally a secret document, it was leaked to the public but ultimately acknowledged by the Discovery Institute. The document outlines a plan to separate science from its "materialistic" moorings and replace it with a science consistent with Christian convictions.

John Farrell, "The Fossils in Our Genes," *Forbes*, October 21, 2011. https://www.forbes.com/sites/johnfarrell/2011/10/21/the -fossils-in-our-genes/?sh=6b3dd8c74b32.

A short article with a link to a scientific paper. Evidence for common ancestry includes "fossil" genes, or pseudogenes. Pseudogenes are nonfunctioning genes in our genomes, once functioning in an ancestor. One such pseudogene is a gene for egg yolk in the human genome.

11

You Can't Make a Monkey Out of Me: The Touchy Topic of Human Evolution

Chapter Highlights

- Human evolution is a touchy topic for many, even if they accept evolution in general.
- A straight-line, one-in-front-of-the-other evolution from chimpanzee to modern human (as depicted in the famous *March of Progress* poster) is a common misconception about human evolution.
- "Lucy" (*Australopithecus afarensis*), a famous transitional in human evolution, is a mosaic of human and chimpanzee traits.
- The *Homo* genus arose in Africa. Some members of *Homo* left Africa about one million years ago and gave rise to human groups closely related to modern humans.
- When modern humans left Africa, they met and had children with other human cousins—the Neanderthals and Denisovans.
- Maps of the human and chimpanzee genomes provide astounding evidence of common ancestry.
- Modern humans descend from an ancestral population of about 10,000 individuals who lived about 150,000 years ago.

Discussion

1. Using what you know about the process of evolution, how would you respond to this statement: "If humans came from monkeys, why do we still have monkeys?"

2. Explain the "cousin-grandmother" analogy regarding human evolution.

3. How is the famous *March of Progress* illustration misleading?

4. Some people of faith are willing to accept evolution and common ancestry for plants and animals, but they draw the line at humans. What do you think motivates this "line in the sand"?

5. Does common ancestry with all animals (and all life) demean human life? Most people do not consider other natural human processes demeaning—for example, human embryotic development and childbirth. How might this inconsistency be explained?

6. Eighty-seven-year-old Mary Adams is the niece of the man who "found" human footprints alongside dinosaur prints in Glen Rose, Texas. Here's Mary: "If we were not created by God, . . . there's no one to whom we are accountable. We can live exactly as we please."[1] Do you agree with Mary? Why or why not? Discuss the implications of this statement.

7. Do you see a conflict between human common ancestry and humans as God's image-bearers? What do you believe about the nature of being created "in God's image"? Is it a material characteristic or a purpose? Is it a resemblance or a vocation?

8. Ancient kings often placed statues (images) of themselves in cities and temples to represent their rule and presence (think about the story of Shadrach, Meshach, and Abed-

nego in the book of Daniel).[2] How might this practice inform our understanding of the "image of God"?

9.   If, as David Menton states, all evolutionist interpretations of human fossils are a combination of imagination and/or deceit, you would expect at least a few nonreligious scientists would agree. But that is not the case: those who reject common ancestry and the human fossil record are always religious. Who is driving the bus regarding human fossils—science or theology? One driver or several?

10. Does the gospel of Christ demand a literal Adam? Does the message of the New Testament fall apart if Adam is not the genetic father of us all? What are the implications of such a position?

Digging Deeper: Are Humans Still Evolving?

Got milk? Chances are you have some in your refrigerator, in one form or another. About 35 percent of the global population includes milk or milk products in their diet. But for the rest of the folks on the planet, it's a big "no thank you."

An enzyme called lactase is needed to digest lactose, the sugar in milk. In the presence of lactase, all is well. Fortunately, babies and young children make lactase— critical for their milk-based diets.

Most adults, however, stop making lactase after childhood. In the absence of lactase, things get ugly. Bacteria in the gut step in and digest the lactose, and the result is bloating, painful cramps, nausea, and diarrhea.

About 35 percent of the planet—mostly of northern and central European descent and some African and Middle Eastern populations—are "lactase persistent." In other words, these folks continue making lactase into adulthood and have no problems drinking milk.

The reason for this lies ten thousand years in the past. It is a story of human evolution.

For most of human history, we were hunters and gatherers. We foraged what we could. We made tools and hunted wild animals. We nursed our babies until they could eat what the adults provided.

At some point a mutation occurred in the lactase gene. The result? Lactase production remained stuck in the "on" position. Only babies were drinking milk, so the mutation went unnoticed.

About ten thousand years ago, we settled down and started farming. And in many parts of the world, we herded animals, including dairy animals.

Around eight thousand years ago in what is now modern Turkey, the lactase-persistence gene increased in frequency. People who carried the gene could drink milk and were likely healthier. Archeological evidence tells us that farmers were making lower-lactose cheese and yogurt, so even lactose-intolerant people could benefit from milk.

As farmers from the Fertile Crescent moved into Europe, they brought their grain crops, and within their genomes, they carried the lactase-persistence gene.

Milk is a close-to-perfect food. It is full of carbohydrates, protein, fat, and calcium. In times of famine, crop failure, and food shortages, milk was a lifesaver to those who could digest it. If milk made you sick, you likely starved.

Quite simply, people who could drink milk were more fit in an age of frequent food insecurity. They were healthier and had more and healthier children. As a result, the gene spread quickly across Europe. Dr. Mark Thomas of the University College London calls the lactase-persistence gene "the most advantageous trait that's evolved in Europeans in the recent past."[3]

Populations in Eastern Africa that have historically raised dairy animals also have high frequencies of lactase persistence. Interestingly, the mutation that keeps lactase turned "on" in the African populations is a different mutation than the one found in Europe. So strong was the advantage of lactose tolerance, the trait evolved multiple times.

Today we can map "lactose hot spots" across the globe. In parts of the world with a history of raising dairy animals, we find high frequencies of lactase persistence. In other parts of the world without a history of dairy farming, lactose tolerance is close to zero.

Where milk was available, people with the gene were at an evolutionary advantage, so the trait took off in the population. Where milk was not available, those with the gene had no advantage. The trait did not increase in the population, and it effectively disappeared.

Lactose tolerance is quite simply evolution by natural selection.

Resources

Dennis R. Venema and Scot McKnight, *Adam and the Genome: Reading Scripture after Genetic Science* (Grand Rapids: Brazos Press, 2017).

> What do you get when a geneticist and a New Testament scholar collaborate on a book about Adam? Excellence. Venema walks us through the evidence of human

population genetics. McKnight follows with a discussion about Adam and Eve in their context, Adam within the Jewish world, and Adam in Paul's writings.

Your Inner Fish, episode 3, "Your Inner Monkey," aired April 23, 2014, on PBS, https://www.pbs.org/video/your-inner -fish-your-inner-monkey/.

Third in a three-part PBS series based on the book *Your Inner Fish* by Neil Shubin, this episode explores the evolutionary history of humans and the traits we inherited from our primate ancestors.

"Great Transitions: The Origin of Humans," HHMI BioInteractive, December 8, 2014, video, 19:44, https://www.bio interactive.org/classroom-resources/great-transitions -origin-humans.

A beautiful short film featuring important fossils from human evolution and interviews with the paleontologists who found them and study them.

Jonathan Tweet, *Grandmother Fish: A Child's First Book of Evolution* (New York: Feiwel & Friends, 2015).

"This is our Grandmother Reptile. . . . She could crawl across the ground. Can you crawl?" This delightful book introduces the idea of common ancestry to young children and includes a beautiful family tree illustrating how all life on earth is related.

12

Leaving Creationism (Without Leaving God)

Chapter Highlights

- As with Galileo in the seventeenth century, modern Christian arguments against science are rooted in theology, not science.
- Despite ancient explanations of weather phenomena in the Bible, we do not read these passages literally. No one insists on biblical meteorology.
- Evolution can be studied and appreciated as the way in which God created life.
- Rejecting science in the name of religion erects barriers to faith.
- A young earth, special creation, and intelligent design all require the natural world to be something it does not appear to be.
- A natural world that is not what it appears has implications about the nature of God.
- Although Genesis is not science or history, it tells us important things about God and creation.

Discussion

1. Galileo was not the only famous astronomer of his day who affirmed a sun-centered solar system. Yet, Galileo

was brought before the church Inquisition. Why was the church unhappy with Galileo? Was the opposition to a sun-centered system scientific or religious? Explain.

2. If we accept evolution, are we believing the ideas of science over the word of God?

3. What aspects of theology do you see underlying opposition to evolution?

4. Were you surprised to hear that "the church's opposition to science" is one of the six primary reasons people decide to leave the church? Why or why not?

5. When science has not fully explained a process or a structure, we are often tempted to put God into the gap. In other words, if science can't provide the answer, it must be a miraculous intervention by God. Do you foresee any problems with this approach?

6. Have we exchanged the idea of an unchanging God who is "the same yesterday, today, and forever" for "we can't apply a twenty-first-century understanding to Scripture"? What might be the consequences of such a conclusion?

7. Most people who change their minds about evolution spend time in transition. Rarely do people go from special creation to evolution and common ancestry overnight. What might a time of transition look like? What struggles might someone in a time of transition encounter?

8. Richard Rohr has written extensively about the role of experience in forming our faith, whether we realize it or not. For example, we experience science in the form of modern medicine, so we don't view seizures as demon possessions, despite biblical example. Our experience influences our interpretation of Scripture. But when it comes to origins, we often encounter a roadblock. We

don't incorporate experience into our interpretation of Scripture. Why might this be so?

9. Can the study of faith and science result in the growth of faith, not the erosion of it? How do you see the way forward toward this goal?

10. How can we be the people of God in a modern scientific age? What discussions regarding applications for faithful living are we missing?

Digging Deeper: Faith in a Secular Age

In Puritan societies of colonial America, gossiping, skipping church, and swearing did not simply land you in hot water with your pastor. These actions were not only sins in God's eyes but were punishable by the civil courts of the day.

For most of the last five hundred years, religion and secular life were virtually indistinguishable. Natural disasters, plagues, and the rule of kings were all seen as acts of God. To declare yourself an atheist would be disastrous in politics, education, or business.

Not so in the twenty-first century. For the first time, unbelief in God is a real option for anyone.

In his classic work *A Secular Age*, Charles Taylor puts it bluntly: "Why was it virtually impossible not to believe in God, in say, 1500 in our Western society, while in 2000 many of us find this not only easy, but even inescapable?"[1]

The answer, according to Taylor, is the rise of the "immanent frame" in the west. The scientific revolution and the Enlightenment birthed a new way of knowing—a knowing based on evidence, empirical data, and reason. Within the immanent frame, everything in the natural world is understandable without reference to anything beyond science.

The immanent frame, says Taylor, supplanted what he calls "transcendence": the world of faith, belief, mystery, and religion.

The Science and Youth Ministry project is a landmark study of modern youth ministry in American churches.[2] The goal of the research was to understand how the modern scientific world impacts teens and the response of youth ministry workers to the issues.

Taylor's frameworks of immanence and transcendence provided a structure for understanding the experiences of teens in American churches. Although the focus of the study was youth ministry, without a doubt, the findings have implications for us all.

None of the teens in the study doubted or challenged the immanence of science. Modern science is their world; it is the air they breathe. Living in a modern world of science and technology does not require belief in God.

Teens are not rejecting faith outright. Teens manage to live in both worlds, as long as the worlds remain separate—faith in church, science in school and everywhere else. But when worlds collide, the teens feel the tension and admit science wins the fight for believability.

Too often the goal of youth ministry is to protect teens from influences that might destroy faith, science included. Teens, on the other hand, are not interested in a protective barrier. Instead, they want to know if it is possible to live in a modern scientific world and maintain belief in God.

And this is where the research findings may surprise you.

Apparently, discussions of science in a faith-based setting prompt teens to examine the plausibility of faith. The conversations went beyond "what does my church say I need to believe" to deeper conversations about the believability of a living and loving God. It wasn't always an easy conversation, and it wasn't always a slam dunk in favor of faith.

Acknowledging the reality of science did not destroy faith but instead prompted conversations about what it means to have faith in a modern scientific world.

Resources

John H. Walton, *The Lost World of Genesis One: Ancient Cosmology and the Origins Debate* (Downers Grove: IVP Academic, 2009).

> A classic! Using original languages and considering culture, Walton defends an understanding of Genesis within its ancient Near Eastern context.

Peter Enns, *The Evolution of Adam: What the Bible Does and Doesn't Say about Human Origins* (Grand Rapids: Brazos Press, 2012).

> Old Testament scholar Pete Enns invites us to read Genesis and the story of Adam with ancient eyes. Enns also includes a discussion of Paul's writings about Adam.

"Galileo—Physicist," *Biography*, YouTube video, 2:51, July 8, 2013, https://youtu.be/2J0-ZbbrD6U.

> Galileo insisted that a hypothesis be tested—you can't just say it's so. This short video summarizes the evi-

dence that convinced Galileo the earth was not the center of the solar system.

Andrew Root, David Wood, and Tony Jones, "Youth Ministry & Science: A Templeton Planning Grant White Paper," Science for the Church, January 2015, https:// scienceforthechurch.org/wp-content/uploads/2020/05 /YouthMinistryAndScience_WhitePaper-1.pdf.

A fascinating study of American youth ministry regarding faith and science. The researchers provide a framework for understanding the discussion from the perspective of teens, the failure of youth pastors to address the issues, and a roadmap for change.

Notes

Chapter 1

1. Ronald L. Numbers, *The Creationists* (Cambridge: Harvard University Press, 1992), 351–52.

2. Numbers, *The Creationists*, 316.

3. Eugenie Scott, "Debates and the Globetrotters," *The Talk-Origins Archive*, July 7, 1994, http://www.talkorigins.org/faqs /debating/globetrotters.html.

4. "Gish Gallop: When People Try to Win Debates by Using Overwhelming Nonsense," *Effectiviology*, https://effectiviology .com/gish-gallop/.

5. "Francis Collins Wins 2020 Templeton Prize," John Templeton Foundation, https://www.templeton.org/news/francis -collins-awarded-2020-templeton-prize.

Chapter 2

1. "The History of Evolutionary Thought," Understanding Evolution, https://evolution.berkeley.edu/the-history-of -evolutionary-thought/1800s/extinctions-georges-cuvier/.

2. Donald R. Prothero, *Evolution: What the Fossils Say and Why It Matters* (New York: Columbia University Press, 2017), 59–61.

Chapter 3

1. Tom Nichols, *The Death of Expertise: The Campaign against Established Knowledge* (New York: Oxford University Press, 2017), 1.

2. "*Staphylococcus Aureus* and Discovery of Penicillin," Bacteria in Photos, http://www.bacteriainphotos.com/Alexander_Fleming_and_penicillin.html.

3. "The Lederberg Experiment," Understanding Evolution, https://evolution.berkeley.edu/the-lederberg-experiment/.

Chapter 4

1. Allan Chapman, *Slaying the Dragons: Destroying Myths in the History of Science and Faith* (Oxford: Lion Books, 2013), 121.

2. Ronald L. Numbers, *The Creationists* (Cambridge: Harvard University Press, 1992), 60.

3. John C. Whitcomb and Henry M. Morris, *The Genesis Flood* (Phillipsburg, NJ: P&R Publishing, 1961).

4. Numbers, *The Creationists*, 60.

5. "How Old Is the Earth?," Answers in Genesis, https://answersingenesis.org/age-of-the-earth/how-old-is-the-earth/; "Can the Ussher Chronology Be Trusted?," Institute for Creation Research, https://www.icr.org/article/can-ussher-chronology-be-trusted/.

Chapter 5

1. "Airport Runway Names Shift with Magnetic Field," National Centers for Environmental Information, https://www.ncei.noaa.gov/news/airport-runway-names-shift-magnetic-field.

Chapter 6

1. *Is Genesis History?* (Nashville: Compass Cinema, 2017), video, 100 min., https://www.youtube.com/watch?v=UM82 qxxskZE.

2. Irving Finkel, "Was the Ark Round? A Babylonian Description Discovered," *The British Museum* (blog), January 24, 2014, https://blog.britishmuseum.org/was-the-ark-round-a -babylonian-description-discovered/.

3. Pete Enns, "Gilgamesh, Atrahasis and the Flood," Bio-Logos, June 1, 2020, https://biologos.org/articles/gilgamesh -atrahasis-and-the-flood.

4. Enns, "Gilgamesh, Atrahasis and the Flood."

5. William Ryan and Walter Pitman, *Noah's Flood: The New Scientific Discoveries about the Event That Changed History* (New York: Simon & Schuster, 1998).

6. Suzanne Trimel, "Ancient Flood Theory Supported by Discovery of Human Artifacts," *Columbia University Record*, September 11, 2000, http://www.columbia.edu/cu/record/ar chives/vol26/vol26_iss2/2602_Flood_Theory.html.

7. Ishaan Tharoor, "Before Noah: Myths of the Flood Are Far Older Than the Bible," *Time*, April 1, 2014, https://time .com/44631/noah-christians-flood-aronofsky/.

Chapter 7

1. Jared Diamond, "Mr. Wallace's Line," *Discover*, August 1, 1997, https://www.discovermagazine.com/planet-earth/mr -wallaces-line.

Chapter 8

1. Robyn Ross, "Tracking Creation in Glen Rose," *Texas Ob-*

server, April 4, 2012, https://www.texasobserver.org/tracking
-creation-in-glen-rose/.

2. Ross, "Tracking Creation in Glen Rose."

3. Sean B. Carroll, "The Day the Mesozoic Died," *Nautilus,*
January 13, 2016, https://nautil.us/the-day-the-mesozoic-died
-rp-8600/.

4. Riley Black, "Fossil Site Reveals How Mammals
Thrived after the Death of Dinosaurs," *Smithsonian Maga-
zine,* October 24, 2019, https://www.smithsonianmag.com
/science-nature/fossil-site-reveals-how-mammals-thrived
-after-death-of-dinosaurs-180973404/.

Chapter 9

1. *Your Inner Fish,* episode 2, "Your Inner Reptile," aired
April 16, 2014, on PBS, https://www.pbs.org/video/your-inner
-fish-program-your-inner-reptile/.

Chapter 10

1. "Malaria and Sickle Cell Anemia," HHMI BioInteractive,
August 26, 2014, YouTube video, 14:15, https://youtu.be/Zsbh
vl2nVNE.

Chapter 11

1. Robyn Ross, "Tracking Creation in Glen Rose," *Texas Ob-
server,* April 4, 2012, https://www.texasobserver.org/tracking
-creation-in-glen-rose/.

2. Dennis R. Venema and Scot McKnight, *Adam and the
Genome: Reading Scripture after Genetic Science* (Grand Rapids: Brazos Press, 2017), 129.

3. "Early Man 'Couldn't Stomach Milk,'" *BBC News,* February 27, 2007, http://news.bbc.co.uk/2/hi/health/6397001.stm.

Chapter 12

1. Charles Taylor, *A Secular Age* (Cambridge: Harvard University Press, 2007).

2. Andrew Root, David Wood, Tony Jones, "Youth Ministry & Science: A Templeton Planning Grant White Paper," Science for the Church, January 2015, https://scienceforthe church.org/wp-content/uploads/2020/05/YouthMinistryAnd Science_WhitePaper-1.pdf.